Chapter 1
ADMINISTRATION

Chapter 2
DEFIN'S

Chapter 4
EMER PLANNING

Chapter 4
EMER PLANNING

Chapter 5
FIRE SERVICE
FEATURES

Chapter 5
FIRE SERVICE
FEATURES

Chapter 6
BLD SERV &
SYSTEMS

Chapter 6
BLD SERV &
SYSTEMS

Section 903
AUTO
SPRINKLER SYS

Section 903
AUTO
SPRINKLER SYS

Section 907
ALARM &
DETECT SYS

Section 907
ALARM &
DETECT SYS

Chapter 10
MEANS OF
EGRESS

Section 1006
NO. OF EXITS

Section 1006
NO. OF EXITS

Chapter 11
EXISTING
BUILDINGS

Chapter 12
ENERGY
SYSTEMS

Chapter 12
ENERGY
SYSTEMS

Chapter 20 AVIATION FACILITIES

Chapter 23
FUEL DISP
& GARAGES

Chapter 24
FLAM FINISHES

Chapter 27
SEMICON FABS

Chapter 27
SEMICON FABS

Chapter 28
LUMBER YARDS

Chapter 29
ORGANIC
COATINGS

Chapter 33
CONST & DEMO

Chapter 33
CONST & DEMO

Chapter 36
MARINAS

Chapter 36
MARINAS

Chapter 40
SPIRITS &
WINE STOR

Chapter 41
TEMP HEAT
& COOK

Chapter 54
CORROSIVES

Chapter 55
CRYO FLUIDS

Chapter 56
EXPLOSIVES

Chapter 57
FLAM &
COMBST LIQ

Chapter 57
FLAM &
COMBST LIQ

Chapter 59
FLAM SOLIDS

Chapter 60
TOXICS

Chapter 66
REACTIVE MATL

Appendix B
FIRE-FLOW
FOR BLDGS

Appendix C
FIRE HYDRANT LOC

Appendix D
FIRE ACCESS ROADS

Appendix D
FIRE ACCESS ROADS